Hetha The Alchemical Wizard With Helium Magic

Element 2 – Magical Elementals Chemistry Fantasy

By Sybrina Durant

Hetha The Alchemical Wizard With Helium Magic

Element 2 – Magical Elementals Chemistry Fantasy

Paperback ISBN: 978-1-942740-71-1

BISAC Codes:
JUV036000 - JUVENILE FICTION / Science & Technology
JUV037000 - JUVENILE FICTION / Science Fiction, Fantasy, Magic
JUV002270 JUVENILE FICTION / Fantasy / Dragons, Unicorns & Mythical Creatures

League City, Texas, United States of America

Contact Sybrina@sybrina.com.

No Metal
Hetha
Hetha The Wizard With The Helium Staff
He
No Magic

Chapter 1: The Crystalline Wand and the Warning Wind

The village of Thornmere sat nestled between two ancient ridgelines in the heart of MarBryn, where copper weathervanes spun atop stone chimneys and alchemical lanterns — glass globes filled with luminescent gas — lined the cobblestone streets in amber rows. It was a place where old-world blacksmiths worked beside new-world pneumatic forges, where horse-drawn carts shared dusty lanes with steam-wheeled carriages, and where the smell of sulfur from the nearby refinery mixed pleasantly with the scent of baked bread from the market square.

It was here, on a cool morning when the mist still clung to the lower ridges like a second skin, that Hetha the Alchemical Wizard descended the crooked steps of her laboratory tower.

She was not a towering figure, nor was she the kind of wizard who announced herself with booming thunderclaps. She was slender and swift-moving, with a cascade of copper-red hair that caught the morning light and threw it back in warm, glowing ribbons. Her skin carried that faint luminescence that made people look twice, as though she had swallowed something magical and it had not quite settled. Which, truthfully, was closer to the truth than most people realized. Her eyes — sharp and vivid blue, like a gas flame burning at its cleanest — scanned the street with the same curiosity they had carried since childhood. She wore her royal blue robes today, gold buttons running in a neat line from collar to hem, a broad golden brooch clasped at her throat. Her sturdy brown boots, decorated with fine gold filigree at the toe and ankle, clicked against the cobblestones as she walked.

Hetha had been like this since she was small — the kind of child who took apart a music box to see what made the chime happen, then put it back together better than before. She had grown into a woman who thought in careful, deliberate steps, who rarely raised her voice because she rarely needed to, and who believed — quietly, stubbornly — that almost any problem could be solved if you understood it precisely enough. She was not reckless, but she was brave in the particular way of someone who trusted her own preparation. She read before she acted. She measured twice. And when she finally moved, she moved with the calm certainty of someone who had already run the numbers in her head three times over.

In her right hand, she carried her wand: a golden rod with a flawless crystal of solidified helium sealed at its tip behind reinforced alchemical glass. The crystal caught the morning light and scattered it in tiny prismatic bursts across the walls of nearby buildings.

Beside her, trotting with the slightly exaggerated dignity of a creature who believed himself very important, was Pim.

Pim was a Flarkin — a species of small, fox-like magical creature native to the northern marshes of MarBryn. He stood roughly knee-height to Hetha, his rust-orange fur interrupted by bold streaks of white along his flanks, and his three bushy tails flicked rhythmically behind him. His amber eyes were large and extraordinarily expressive. Flarkins were known across MarBryn for two things: their loyalty, which was total and often inconveniently dramatic, and their tails, which stored small charges of static electricity that built up naturally in their thick fur the way a wool blanket gathers a spark on a dry winter night. Pim had been Hetha's companion since he was barely old enough to walk on all four legs at once, and in all that time he had developed a personality that could only be described as a professor who had never quite gotten his degree — full of opinions, mostly correct, delivered with unearned confidence and a genuine, if occasionally exhausting, devotion to Hetha's safety.

"You forgot your compendium again," Pim said, in a voice that was smooth and slightly theatrical, as though he had once studied at an academy for formal speech.

Hetha patted the satchel at her hip. "I did not. I moved it to my satchel last night."

"Ah." Pim paused. "I was hoping to feel useful this morning."

"You can feel useful by telling me whether you noticed anything strange on the ridge last night. I heard howling that wasn't wolves."

Pim's three tails went still — a sure sign of unease in any Flarkin. "I did notice something," he admitted. "Around the second bell. A light, high on the Greyvane Ridge. The color was wrong. Too green. And it moved against the wind."

Hetha slowed her pace. The market square was waking up around them — a merchant was loading pressurized canisters of alchemical gas onto a cart, their identification runes stamped in red, while a pair of apprentice chemists in leather aprons argued over a brass instrument neither of them could calibrate. The ordinary life of Thornmere proceeded as it always did.

But the green light troubled her.

She unclasped her satchel and drew out the compendium — a thick volume bound in deep green leather, its spine reinforced with brass rivets. The pages were a hybrid of ancient parchment and modern chemical notation, filled with her own cramped handwriting alongside printed diagrams of molecular structures and arcane formations. She flipped to a section near the back, running her finger along an annotated margin.

"A green displaced light moving against wind currents," she murmured. "That matches the signature of forced elemental disruption. Someone interfered with an atmospheric node last night, Pim."

"Deliberately?"

"The wind doesn't make mistakes," she said grimly.

She looked up at Greyvane Ridge, where the morning mist was already beginning to thin — and where, she now noticed, a single black bird circled in a patient, waiting spiral.

"Pack your things tonight," she told Pim quietly. "We leave at dawn."

Chapter 2: The Alchemist's Road and a Stranger Named Corvin

The road north from Thornmere was called the Alchemist's Road, and it had earned its name honestly.

Along its thirty-mile stretch toward the mountain city of Caldervane, travelers passed sulfur springs that bubbled at the roadside like natural laboratories, deposits of raw mineral crystal jutting from exposed rock faces in colors that ranged from deep amber to electric violet, and Iron mile-markers engraved not with numbers but with elemental symbols — a tradition begun by the old Guild of Wandering Chemists two centuries prior. The road itself had been treated at its founding with a mineral sealant developed by a long-dead alchemist named Wreth Dunmore, and it had not cracked significantly since.

Hetha and Pim had been walking since dawn, the mist burning off the ridges around them as the sun climbed. Hetha's satchel was heavy with the compendium, a sealed vial of reagents she had prepared the evening before, and a small collapsible pneumatic scope — a new-world device of brass and glass she had purchased from a traveling instrument maker three months ago. Pim trotted at her side, his three tails swaying in easy rhythm.

By midmorning, hunger had caught up with both of them, and Hetha called a halt at a flat outcrop overlooking the valley, where a stunted pine offered a thin ribbon of shade.

She unpacked their meal with the same economy she brought to everything else: a small clay pot of spiced lentil stew, kept warm since dawn by a charm no more complicated than a well-sealed lid and a stone that held heat; a wedge of hard, sharp cheese wrapped in waxed cloth; and a flask of rosehip tea gone lukewarm but still fragrant. She ate while she read, propping the compendium open against her knee, alternating a spoonful of stew with a line of notation, as though food and study were simply two forms of the same nourishment.

Pim, by contrast, treated mealtime as an event deserving of its full attention. He produced, with evident satisfaction, a small oilcloth parcel of honeyed marsh-nuts and a strip of smoked river fish — his two great weaknesses, and the subject of more of his conversation than Hetha would have liked to admit.

"You could at least pretend to enjoy your stew the way I enjoy this fish," he said, delicately tearing off a piece.

"I enjoy my stew a great deal," Hetha said, not looking up from the compendium. "I simply do not narrate the enjoyment."

"A tragic waste of a good meal," Pim said, and ate another honeyed nut with theatrical relish. Flarkins, as a species, ran hot and burned through their reserves quickly — three tails' worth of stored static charge did not maintain itself for free — and Pim genuinely needed the fat and sugar he loved so much, a fact he was always pleased to cite as justification.

"Tell me about the atmospheric nodes," Pim said. He had an exceptional habit of demanding education while walking.

"MarBryn has seven primary atmospheric nodes," Hetha began, settling into the comfortable cadence of explanation. "They regulate the flow of elemental energy across the land. Think of them as valves in a great gas pipeline — each one controls pressure and distribution. The Greyvane node sits on the ridge above Thornmere and governs the northern air corridor. If someone forced it open or blocked it artificially, the effects would ripple outward."

"And the green light?"

"Green elemental displacement is associated with volatile compound magic. Someone used reactive chemistry — forced alchemical synthesis — at the node site." She paused. "That's not accidental work. That's deliberate and dangerous."

Pim made a sound somewhere between a growl and a worried hum. "Who would do that?"

Hetha's jaw tightened slightly. "I have a suspicion. But I need more evidence before I say the name."

They walked in thoughtful silence for a while. The road curved around a limestone bluff, and as they rounded it, they nearly collided with a figure sitting on the roadside milestone, eating an apple with the relaxed air of someone who had nowhere urgently to be.

He was young — perhaps twenty-five — with a lean, angular face framed by close-cropped dark hair and a jaw dusted with three days of stubble. He wore the practical clothing of a field engineer: reinforced canvas trousers, a belted jacket with multiple pockets, and heavy-duty gloves tucked into his belt. A large rucksack leaned against the milestone beside him, bristling with equipment — coiled tubing, brass fittings, pressure gauges, and what appeared to be a compact gas analysis unit. He had the unhurried economy of movement of someone who trusted his tools completely and rarely wasted effort on anything that wasn't load-bearing. Later, Hetha would learn that Corvin Strake ate the way he worked — practically, without fuss, mostly out of a field ration tin, though he would go visibly quiet with pleasure over a piece of candied ginger, a small vice he pretended not to have.

"You're Hetha," he said without looking up from his apple. "The Helium Wizard from Thornmere."

"I am," she said, measuring him carefully with those blue eyes. "And you are?"

"Corvin Strake." He finally looked up, revealing gray eyes that were sharply intelligent and faintly amused. "Field engineer. Independent contractor. I've been hired by the Caldervane Elemental Survey Office to assess disturbances along the northern node corridor." He reached into his breast pocket and produced a folded authorization paper, stamped with the Survey Office's brass seal.

Hetha examined it and handed it back. "What did your instruments show last night?"

Corvin raised an eyebrow. "You felt it in Thornmere?"

"We saw it."

He stood, folding the apple core into a small waste pouch and shouldering his rucksack. "My portable gas analyzer lit up around the second bell. Elevated concentrations of volatile compound residue in the air column above Greyvane Ridge. Specifically — " he pulled a small data scroll from his jacket and unrolled it — "an unusual tri-molecular signature I've never seen in natural readings."

Hetha leaned over the scroll. Her eyes sharpened. "That's not natural at all. That's manufactured."

"Exactly my conclusion." He looked at her steadily. "I'd rather not walk into Caldervane alone with that news. The city council has a tendency to dismiss field engineers under thirty." He glanced at her wand and the glinting crystal at its tip. "An Alchemical Wizard of your reputation would be considerably harder to dismiss."

Pim looked up at Corvin with large, assessing amber eyes. "I like him," he announced. "He has calibrated instruments and a healthy distrust of bureaucracy."

Hetha allowed herself a small smile. "Walk with us, Corvin Strake."

Chapter 3: Caldervane and the Council of Pressures

Caldervane was magnificent in the way that only a city built into the face of a mountain could be.

Its lower districts occupied a broad natural terrace of grey-green stone, where the buildings were thick-walled and medieval in spirit — great halls of cut limestone, tower-houses with narrow lancet windows, bridges of ancient ironwork spanning the deep channels carved by underground rivers long since redirected by alchemical engineering. But above the medieval base, the city climbed in layers of increasing modernity. Mid-level districts housed the pneumatic rail terminus, the pressurized gas works — a forest of copper pipes and gleaming condensation towers — and the famous Caldervane Academy of Applied Chemistry, its long windows glowing amber at night with the light of ongoing experiments. At the summit, carved directly into the mountain's peak, sat the Hall of Elemental Governance, its façade decorated with enormous relief sculptures of the periodic table rendered in polished obsidian tiles.

Hetha had always loved Caldervane and found it slightly overwhelming in equal measure.

They arrived at the lower gate in the mid-afternoon. The gate itself was a marvel — two enormous doors of reinforced steel and alchemical glass, etched with elemental formulas, operated by a pneumatic mechanism that hissed and sighed as it admitted them. A customs officer in a white coat noted their names and wand registration number, ran a spectral analysis on Corvin's gas unit to verify it was not a prohibited device, and waved them through.

Inside, Pim drew looks. Flarkins were uncommon in the city.

"People keep staring at my tails," he muttered.

"You have three of them," Corvin said reasonably.

"That is simply efficient."

The Survey Office received them in a paneled room on the third floor of an administrative building that smelled of chemical solvents and old paper. Three officials sat behind a curved desk — Director Maren Solley, a brisk woman in her fifties with silver-streaked hair, sharp reading spectacles perched permanently on the end of her nose, and the economical manner of someone who had processed a great deal of bad news in her career and had long since stopped wasting words on anything but the necessary ones; her deputy, a tall man named Foss who took notes in a compact, pressurized ink pen and rarely spoke unless directly addressed; and a junior analyst who appeared to be approximately nineteen and deeply frightened, his hands twisting the corner of a report he had clearly reread a dozen times waiting for this meeting.

Hetha presented the compendium's relevant entries. Corvin laid out his data scrolls. Together, they built a picture.

"The Greyvane node was forcibly excited," Hetha said, tracing the analysis with her finger. "The tri-molecular compound residue Corvin's instruments detected is consistent with a catalytic agent being introduced into the node site. Not a natural occurrence — a deliberate one. The person who did this understands elemental node chemistry at an advanced level."

Director Solley folded her hands. "You believe this is sabotage."

"I believe this is preparation," Hetha said carefully. "A node excitation alone doesn't accomplish much. But if the same was done to the Caldervane node — which sits three miles above this city — the pressure differential between the two disturbed nodes would generate an atmospheric shockwave. Localized. Devastating."

The room was quiet for a long moment.

"Who would do this?" Foss asked.

Hetha exchanged a glance with Pim, who gave a small, deliberate nod.

"I believe it is the work of Mordavek," she said. The name fell into the room like a stone into still water.

Director Solley's expression hardened. "The Corrosion Mage. We believed him contained in the Ashfen Wastes."

"He was contained," Hetha said. "He is no longer."

Mordavek — real name unknown, the title earned for his mastery of destructive chemical decomposition magic — had been one of MarBryn's most feared rogue alchemist for two decades. He had dissolved the foundational alloy of the Westbridge Dam with a compound of his own creation, flooding three villages. He had corrupted the water supply of Dunhallow with reactive mineral agents. He was patient, methodical, and possessed of a profound contempt for everything stable and flourishing.

He had also, Hetha knew personally, once tried to destroy her mentor's laboratory and nearly succeeded.

"We need to reach the Caldervane node before he does," she said, standing. "Who can guide us up the summit road?"

The frightened junior analyst raised his hand. "I can," he said. "I mapped it last season. My name is Tallo."

Pim looked at him with those enormous amber eyes. "Welcome to the company, Tallo," he said solemnly.

Chapter 4: The Summit Road and Floating Stones

The summit road was not, strictly speaking, a road.

It had begun as one — the Survey Office's maps showed a planned pneumatic-assisted lift track that was supposed to carry equipment and personnel up Caldervane Peak's western face. Funding had run out two years prior, leaving the project at roughly forty percent completion. What remained was a partially constructed rail bed, a series of automated gas-pressure pumping stations abandoned mid-build, and a hiking trail of considerable ambition and uncertain maintenance.

Tallo, it turned out, was not merely a junior analyst who had drawn maps. He was a wiry young man of twenty-two with ink-stained fingers, an encyclopedic knowledge of the mountain's geological strata, and an endearing habit of narrating everything he observed as though composing a field report. He had the kind of restless, upward-tilting energy of someone who had spent his whole life being told to be quieter and had simply decided, at some point, that the mountain was worth talking to even if the people around him occasionally weren't listening. Beneath the nervous chatter was a steady, capable person — the kind who noticed the exact width of a crack in stone, or the precise moment weather was about to turn, well before anyone else did.

"We're presently crossing a band of calcium carbonate — limestone of the Deveric period, note the fossils — quite beautiful — transitioning at the next switchback into a metamorphic — "

"Tallo," Corvin said patiently, "can you narrate and walk faster simultaneously?"

"Absolutely yes." Tallo accelerated without interrupting his geological monologue.

They were two hours up the summit road when they encountered the first obstacle.

The trail rounded a cliff face and opened onto a wide ledge — and the ledge was wrong. A section of it, roughly forty feet across, had been destabilized. The rock was riddled with fissures in a pattern Hetha recognized immediately: reactive acid compound erosion. Someone had treated the stone with a corrosive agent, timed to degrade under foot pressure.

"Mordavek's work," she said, crouching to examine the fissure pattern without touching it. "He's set a trap on the approach. Step on this section with any real weight and the whole ledge gives way."

"Can we go around?" Corvin asked.

Tallo was already consulting his map. "There's a secondary path — but it adds four hours and brings us to the node site from below rather than above. We'd lose the positional advantage."

Hetha straightened and considered the ledge. Forty feet of compromised stone, a sheer drop to the left, solid cliff to the right. She looked at her wand, at the helium crystal burning with its cold internal light.

"I can take us across," she said. "One at a time."

She turned to Pim. "Do you trust me?"

"Unconditionally," he said. "Though I reserve the right to make extremely undignified sounds."

Hetha extended her wand toward Pim, channeling her focus through the crystal. She called this technique the Levity Draft — the first spell her mentor had ever taught her, and still the one she trusted most completely. In the real world, helium is roughly seven times less dense than the air around it; a balloon filled with it rises for the same reason a piece of wood floats on water — it simply weighs less than the space it displaces. The Levity Draft did the same thing to a person: Hetha's magic didn't make Pim weightless so much as it convinced the air beneath him to behave, briefly, like something far lighter than itself, so that ordinary buoyancy did the rest of the work. The Helium magic responded like a breath drawn in — a smooth, clean pull of energy. She felt the familiar sensation of density shifting under her intent, and Pim rose from the ground with a startled but dignified yelp, floating to precisely chest height and remaining there, stable and perfectly buoyant.

"Oh," he said, in a small voice. "That's rather nice, actually."

She walked him across the ledge first, setting him down gently on the solid ground beyond. Then Corvin — who gripped the straps of his rucksack tightly but managed to maintain an expression of engineering calm throughout. Then Tallo, who narrated his own levitation with breathless enthusiasm.

Hetha crossed last, barely touching the compromised rock as she stepped quickly across, keeping her center of gravity reduced with a partial levitation she applied to herself — a technique that required significant concentration, as manipulating her own buoyancy was always more difficult than working on others.

On the far side, she let out a slow breath.

"He anticipated us," she said. "Or anticipated someone."

"He anticipates everything," Pim said quietly. His tails were still again. "I've been thinking about what you said — about the atmospheric shockwave. What would it actually do to Caldervane?"

Hetha looked down at the city below them, its copper pipes and lantern-lit windows glittering in the afternoon sun. "The over-pressure wave alone would shatter every alchemical glass structure in the lower city. The gas works, the academy laboratories — pressurized systems failing simultaneously." She paused. "Hundreds of people, Pim."

The silence on the ledge was absolute except for the wind.

"Then we don't let him reach the node," Corvin said flatly.

"No," Hetha agreed. "We do not."

Chapter 5: The Node Chamber and a New Ally

The Caldervane atmospheric node was housed in a structure that had not been built by any engineering team.

It was entirely natural in origin — a cave system at the mountain's peak, where ancient geological pressure had forced a convergence of mineral crystal formations so dense and so precisely arrayed that they had, over the course of ten thousand years, become a functional elemental regulator. MarBryn's old scholars had discovered it, mapped it, and added only minimal interventions: a measured series of observation alcoves cut into the rock, brass instrument panels connected to the Survey Office's monitoring network by wire cables that snaked down the mountain's interior, and a sealed door of reinforced steel and alchemical glass at the entrance.

The door was already open when they arrived.

The four of them stopped at the threshold. Hetha held up her wand, the helium crystal casting its clean prismatic light into the cave's interior. The crystals within — some the size of a fist, others the size of a cart — glowed in response, deep violet and pale blue, and the air inside had the slightly electric quality of a space where elemental energy was dense.

But something else was inside. A figure moved among the crystals — small, quick, not human.

"Stop," Hetha said, in a voice that carried without shouting. "Identify yourself."

The figure stopped. Then, slowly, stepped into the wand-light.

It was a Prism Sprite — one of the rare elemental beings that existed only in environments of high crystal density. She stood roughly two feet tall, her body semi-translucent, her form humanoid but edged in light like a figure cut from clear glass. Her features were fine and sharp, her eyes enormous pools of shifting spectral color. She moved with a strange, gliding economy, as though gravity applied to her only as a courtesy, and her voice, when it came, arrived a half-second alongside a faint chiming echo, like a bell struck under water.

"I am Lys," she said. Her voice had a bell-like harmonic quality, each word accompanied by a faint resonance. "I am the keeper of this node. I have been here for a very long time."

"Has anyone else entered recently?" Hetha asked, stepping carefully inside.

"Three nights ago," Lys said, her spectral eyes troubled. "A man with an iron casing on his left arm and dark compound magic. He planted something in the secondary formation — " she pointed with a finger that refracted light as it moved — "there. Behind the second cluster."

Corvin was already moving toward it with his instruments. He crouched behind the indicated crystal formation and shone a small chemical light into the gap. "There it is," he said grimly. "A catalytic trigger device. Old-world ceramic housing, new-world compound charge. If this goes off — "

"It matches the Greyvane excitation signature," Hetha said, joining him. She examined it without touching, using her scope. "He needs both nodes triggered simultaneously to generate the shockwave. This is the second charge. The Greyvane device is already active."

"Can you disarm it?" Pim asked.

Hetha sat back on her heels and opened the compendium. She found the relevant section and read with the focused speed of someone who had studied this exact type of compound construction before. After a long moment, she looked up.

"The trigger mechanism is pressure-sensitive. If I increase the weight on it above its activation threshold, it will lock — not detonate. It needs upward pressure to fire."

She looked at her wand.

"I can make it heavier," she said slowly. "Much heavier. Immovable. And if the weight presses down rather than allowing the upward pressure mechanism to release — it disables permanently."

"You're certain?" Corvin asked.

"Reasonably certain," she said. "Which is the best certainty available."

She extended her wand carefully, drawing on the inverse of her usual power — not lightness and float, but weight and press. She called this counter-spell the Anchor Press, and it worked on the same real principle as the Levity Draft, only run in reverse: just as compressing a gas into a smaller space raises its density and its pressure, Hetha's magic could gather helium's density into a tight, heavy column instead of a buoyant one, weighing down whatever it surrounded the way a diver's tank feels heavier the deeper it sinks. The helium crystal worked both directions: buoyancy added or removed, density raised or reduced. She channeled the force with controlled breath, pushing density into the catalytic device, feeling the mechanism strain and then — with a sound like a very small, very defeated sigh — lock solid.

The glow of the crystal formations stabilized. Lys exhaled with a resonant ring.

"It is safe," Lys confirmed, her spectral eyes reading the node energy. "For now."

"But Greyvane is still active," Tallo said.

"Yes," Hetha said, standing. "And we need to return there before Mordavek does."

Lys looked at Hetha steadily, her translucent form shimmering in the crystal light. "I will come with you," she said. "This node is my home. But the Greyvane node — and all the nodes — are my concern. I am the last of my kind in the northern range. I have nowhere else to be."

Pim's three tails rose in a pleased wave. "Then you are very welcome," he said warmly.

Chapter 6: Shadows in the Ashfen Wastes

The most direct route back to Greyvane Ridge did not pass through Caldervane.

Tallo's maps — spread across the floor of a roadside shelter where they stopped to plan — showed a route through the eastern lowlands that would cut nearly a day off the journey: through the Ashfen Wastes, a desolate stretch of marshland and mineral flat where the soil was saturated with chemical residue from ancient industrial processing. Nothing grew in the Ashfen Wastes that hadn't adapted aggressively to sulfurous ground and acidic air. Everything there was either very hardy or very dangerous.

"Mordavek came from the Ashfen Wastes," Hetha said. "He knows this terrain better than anyone."

"Which means he could be tracking us through it rather than us tracking him," Corvin said, his gray eyes steady.

"Yes. We should assume that." She looked at the company — Pim with his alert amber eyes, Corvin checking his instruments with practiced efficiency, Tallo tracing the map with an ink-stained finger, Lys sitting lightly at the shelter's edge, her translucent form dimmer in the grey light outside the crystal-rich mountain.

"Everyone stay close and watch your chemical exposure. The ground gases in the Ashfenare real hazards. Corvin, you have your analyzer?"

"Always." He patted his jacket. "I'll monitor continuously."

They set out at first light, and the Ashfen Wastes received them with the particular coldness of a landscape that had never been hospitable to anything.

The ground was pale grey-white, crusted with mineral deposits that crunched underfoot and occasionally exhaled thin wisps of sulfur-tinted vapor. The vegetation — where it existed — was skeletal and grey-barked, the leaves narrow and waxy against chemical exposure. The sky above was clear but somehow colorless, as though the air itself had been bleached.

They were four hours in when Corvin's analyzer chirped a warning.

"Elevated concentration ahead," he said quietly, reading the display. "Not sulfur — something more complex. This is a compound mixture. It's been — " he recalibrated, frowned — "it's been dispersed deliberately. Someone seeded this area."

"An ambush preparation," Pim said.

They came from the flats — six of them. Mordavek's agents, wearing respirator masks of hammered iron and wielding weapon-staves tipped with concentrated acid vials, emerging from the mineral crust where they had been lying nearly flat and invisible.

"Scatter left!" Hetha called.

The company broke apart with practiced speed. Hetha raised her wand and sent a pulse of buoyancy force outward — a wide-area version of the Levity Draft she privately thought of as the Gale Lift — a wide-area levitation blast that caught two of the attackers and flung them spinning upward, their staves clattering free. Pim, faster than anything fox-sized had the right to be, had already circled behind the nearest attacker, his triple tails whipping out and striking the weapon-staff from the man's grip with a crack of stored elemental charge. He called the move a Static Snap, and though it had nothing to do with helium, it was every bit as rooted in real science: a Flarkin's fur builds up a static charge from friction the same way a wool sock does crossing a carpet, and Pim had simply learned, over years of practice, to release that charge in a single controlled burst instead of letting it dissipate harmlessly — a Flarkin's natural defense, a sharp static discharge from the tail tips.

Corvin had not drawn a weapon. He had instead opened the front panel of his rucksack and activated a portable pressure-pulse device — an industrial tool meant for clearing blocked pipelines, repurposed with some unauthorized modifications into a very effective blunt-force projector. It sent a compressed burst of air that knocked two attackers flat simultaneously, no magic at all, only the same pneumatic principle that ran half the machines in Thornmere.

Tallo, who had no magic and no combat modifications, threw his geology hammer with exceptional accuracy and took out the respirator mask of the sixth attacker, who immediately succumbed to his own ambient chemical cloud and sat down very suddenly, coughing.

Lys moved among the crystals of her own structure, drawing in light and releasing it in disorienting spectral bursts that scrambled vision and balance — what she called, when Hetha asked her about it later, a Refractive Bloom: real prisms split ordinary light into every color at once, and Lys had simply learned to do it violently and all in one place, overwhelming an eye that expected steady, single-colored light with more information than it could process in an instant.

The ambush collapsed in under two minutes.

"Efficient," Pim said, surveying the scene with his tails fanned.

Hetha crouched beside one of the unconscious attackers and examined the iron-stamped insignia on his respirator mask. "Mordavek's circle," she confirmed. "He knows we're coming."

"Then we move faster," Corvin said.

Chapter 7: The Greyvane Node at Night

They reached the base of Greyvane Ridge by evening, emerging from the Ashfen Wastes into the comparatively welcoming pine-dense foothills that stepped upward toward Thornmere. The familiar scent of the lower ridge — pine resin and cool mineral water from the Greyvane spring — was sharp and grounding after the chemical desolation of the Wastes.

But none of them relaxed. Mordavek was ahead of them, somewhere on the ridge.

Hetha paused at the edge of the tree line and pulled out the pneumatic scope, training it up the slope. Through its brass-ringed lens, she could see the node entrance — a natural stone arch at the ridge's crown, just as it appeared on the Survey Office's charts. And there, at the arch's base, a cold green light.

"He's there," she said quietly.

Corvin took the scope when she passed it. His jaw tightened. "He's already working on the node. If the Caldervane device is locked, what is he actually doing?"

"He can re-trigger the Greyvane device manually without needing the dual activation," Lys said. Her form brightened slightly in the gathering dark, her spectral light rising instinctively. "A single-node excitation will not produce the city-killing shockwave — but it will destabilize the entire northern atmospheric corridor. Storms, pressure failures, elemental cascade across the network." She paused. "It would still be catastrophic. Slower, but catastrophic."

"We go up now," Hetha said. "In the dark. That's our advantage — his light announces him. We come without light if possible."

Lys dimmed herself deliberately, a control that took visible effort but was accomplished quietly. Corvin switched off his analyzer's display light. Pim's tails settled low and close, reducing his visible outline.

The ascent in darkness was careful and slow. The ridge trail was familiar to Hetha — she had walked it dozens of times from Thornmere — but familiarity didn't make the footing less precarious at night. They climbed in single file, stepping by feel and memory, communicating in gestures and whispers.

They were fifty yards from the arch when Hetha stopped them.

Mordavek stood at the node entrance, and she could see him clearly now without the scope's mediation. He was tall and lean, his dark robes trimmed with the dull copper-green of corroded metal — a deliberate affectation, that color, chosen from his element's character. His left arm bore an iron mechanical casing from wrist to elbow: a compound-delivery system of his own engineering, capable of synthesizing and releasing corrosive agents in focused streams. His face was angular and deeply weathered, his eyes carrying that particular deadness of someone who had long since decided that destruction was a reasonable solution to inconvenience. There was, Hetha thought, watching him from the dark, something almost tragic in how precisely he moved — the same discipline she prized in herself, aimed entirely at breaking things instead of understanding them.

He had not heard them yet.

"I can disable the trigger device if I get close enough," Hetha breathed. "But he has to be distracted."

"I can give you sound from the far side of the arch," Lys murmured. "Spectral resonance — it will appear to come from behind him." She called this trick the Prism Throw, a cousin of the Refractive Bloom: instead of splitting light, she bent sound the way a curved mirror bends light, sending it bouncing off the crystal formations to land somewhere she wasn't.

"I'll back her up with a pressure burst once you're in position," Corvin said.

Hetha nodded, gripped her wand, and began to move.

Chapter 8: The Battle at the Threshold

The Prism Throw Lys generated was extraordinary — a deep, resonant thrumming that seemed to rise from the rock itself, directionally perfect, pulling Mordavek's head around toward the dark far side of the arch.

It gave Hetha three seconds.

She broke from cover at a controlled run, crossed the final distance to the node entrance, found the trigger device mounted inside the first crystal formation — a mirror of the Caldervane device, same ceramic housing — and drove a full-density Anchor Press into it with her wand in one sustained, focused burst.

The device groaned and locked.

Mordavek spun.

"The Helium Witch," he said, and his voice had the flat, unhurried quality of genuine contempt. "I wondered when you'd arrive."

"Mordavek." Hetha kept her wand raised, the crystal blazing cold white against the green of his compound light. "It ends here."

"Does it?" He raised his iron arm and the casing hissed, compound agent collecting at the release port. "I've been patient for twenty years. The nodes, the cities, the orderly little civilization MarBryn is so proud of — it's all built on chemical lies. Unstable systems dressed in pretty lantern light. I am merely accelerating the inevitable."

"That's a long speech for someone who set a trap on a mountain path and sent six masked thugs through a swamp," Corvin said from behind him.

Mordavek turned and fired the compound stream in one clean motion.

And then everything happened very fast.

Corvin dove left, rolling across the stone, the compound stream hissing past him and eating into the rock where it landed. Pim launched from the ridge edge in a full-speed leap, three tails crackling with static discharge, and struck Mordavek's iron arm with all three simultaneously in a Static Snap — the electric burst was small but disrupted the delivery mechanism's calibration, sending the next stream wide.

Mordavek swept his arm around and Pim tumbled, striking the rock hard, letting out a sharp yelp.

Hetha felt something cold and sharp inside her and drew on every grain of her Helium mastery. She pointed her wand directly at Mordavek and reversed the field entirely — not lightness or weight applied to an object, but a column of buoyancy force concentrated around him, what she would later name the Column Lift: the same principle as filling a great balloon with helium and tethering it to nothing, so that it simply rises until something stops it. She poured that same relentless upward pressure around Mordavek's whole body, a dense upward pressure that lifted him bodily from the ground, his iron arm useless at that angle, compound stream firing harmlessly upward into the night sky.

He hung six feet in the air, struggling with the complete and total loss of leverage.

"I've applied approximately three hundred pounds of lifting force to your center of mass," Hetha said, breathing hard but steady. "The helium crystal in my wand can sustain that for considerably longer than you can manage without a surface."

Lys had already moved to Pim, her spectral hands — which carried a gentle, temperature-regulating warmth — checking his injuries. "Bruised, not broken," she reported, her harmonic voice reassuring.

Tallo, with the calm practicality of someone who had spent two seasons doing geological fieldwork alone, had produced a length of high-tensile coring rope from his pack and was already fashioning a series of precise knots with the efficiency of a man who had tied down equipment in mountain storms. He worked without hurry, without drama, and when he was ready, he looked up at Hetha with an expression of simple readiness.

"Whenever you bring him down," Tallo said.

Hetha lowered Mordavek in a slow, controlled descent, keeping the buoyancy field engaged until the precise moment Tallo and Corvin had his arms secured. The instant his boots touched stone, both men moved — Corvin pinning the iron mechanical arm against Mordavek's back at an angle that made it mechanically impossible to aim, while Tallo completed three wrapping loops of coring rope around the arm's casing with practiced speed.

Mordavek did not struggle dramatically. He went still with the contained fury of someone filing everything away for later.

"You've delayed nothing," he said quietly.

"We've locked two of your devices and secured your delivery arm," Hetha replied, lowering her wand but not putting it away. "I'd call that more than a delay."

His weathered eyes found hers, and for just a moment something crossed them that was not contempt — something older and more complicated. "The nodes are still unstable," he said. "My compound agents are already metabolizing in the crystal matrices of both sites. Locking the trigger devices doesn't clean the contamination. In seventy-two hours, the matrices will reach saturation and the excitation will occur passively. Without any trigger at all."

The ridge went quiet except for the wind.

Hetha held his gaze and said nothing, because she knew, with the cold certainty of someone who had read the relevant chemistry in her compendium more than once, that he was telling the truth.

Pim had risen, limping slightly but standing steadily on all four legs, his amber eyes fixed on Mordavek. "Is that true?" he asked Hetha directly.

"It could be," she said carefully. "Catalytic compound agents introduced into a crystal matrix can reach self-sustaining reaction thresholds without a mechanical trigger. It depends on the specific compound and the matrix composition." She looked at Mordavek. "What did you use?"

He smiled, and the smile was the worst thing about him — genuinely pleased. "Ask your compendium."

Corvin's jaw worked. "We need to get him to the Survey Office and then get back to both node sites. If what he's saying is correct — "

"It is correct," Lys said. Her spectral form had brightened again, and her enormous shifting eyes were focused somewhere past all of them, reading something the others could not see. "I can feel the Caldervane node from here. Even at this distance. There is a warmth in its matrix that should not be there. It is slow, but it is growing."

Hetha opened her compendium, standing on the windswept ridge with Mordavek secured behind her and the night pressing in from all directions. She turned pages by wand-light, reading fast, cross-referencing the tri-molecular signature from Corvin's data scroll against the compound agent classifications in the back index.

"Tri-phase reactive mineral compound," she said finally. "Self-propagating above saturation threshold. Neutralized by — " she read, re-read, looked up — "by an elemental counter-agent introduced at the molecular bonding sites. Specifically, a noble gas flush delivered at sufficient pressure to displace the reactive compound from the crystal matrix."

The group stared at her.

Corvin spoke first. "A noble gas flush. Like — "

"Like helium," she said. "Pumped through the matrix at concentration and pressure sufficient to physically displace the compound molecules from the crystal bonding sites." She looked at her wand, at the crystal tip glowing at its end. "My magic manipulates helium's buoyancy and density properties. But what I've never attempted — " she paused, and the honesty of the difficulty showed plainly in her face — "is working at molecular scale. Projecting helium displacement through a crystal matrix rather than applying it to a visible object."

"Can you do it?" Pim asked.

She closed the compendium slowly. "I don't know," she said. "But I'm going to try."

Mordavek, hands bound behind him, watched her with an expression that had shifted — almost imperceptibly, almost against his will — from contempt to something approaching curiosity.

"Seventy-two hours," he said again, as they began the descent.

"I heard you the first time," Hetha said.

Chapter 9: The Molecular Attempt

They delivered Mordavek to the Caldervane Survey Office at dawn, turning him over to the city's elemental enforcement officers — a unit that operated out of a reinforced sub-level of the Hall of Elemental Governance and had been waiting for Mordavek's recapture for eleven years. Director Solley was present for the handover, her expression carrying the controlled relief of someone who had not allowed herself to hope too openly and was now quietly recalibrating.

"You have seventy-two hours," Hetha told her. "Or less. I need the Academy of Applied Chemistry's node interface laboratory opened immediately. And I need access to the mountain's internal pipe network."

Solley did not ask unnecessary questions. Within an hour, Hetha had what she needed.

The Academy's node interface laboratory occupied the highest floor of the Academy building, directly below the mountain's upper chambers. It was a long room of stone and steel, its walls lined with pressure gauges, its ceiling threaded with copper pipe connections running to the mountain's interior infrastructure. In the center stood the primary interface console — a combination of old-world alchemy workbench and new-world instrumentation, brass fittings alongside digital pressure readers, crystal observation chambers beside calibrated flow meters.

It was exactly what Hetha needed, and it terrified her slightly.

"Walk me through the theory again," Corvin said, standing at the secondary console with his sleeves rolled up and his gas analysis unit connected to the interface network. Tallo was at the pipe junction board, his geological knowledge of the mountain's interior proving unexpectedly invaluable for identifying the correct routing channels.

Lys stood within the primary crystal observation chamber — a glass-sided enclosure containing a sample of the node matrix crystal — her translucent form pressed gently against the crystal, reading its energy state continuously.

"The compound agent is bound to the crystal matrix at the molecular bonding sites," Hetha said, standing at the central bench with the compendium open and her wand resting beside it. "Helium is a noble gas — it doesn't bond with other compounds under normal conditions. At sufficient pressure and concentration, I can use my magic to project helium displacement force at a scale small enough to act at the molecular interface. The helium molecules physically push the reactive compound molecules out of the bonding sites. The compound becomes unanchored, disperses, and the reaction cascade cannot reach saturation."

"You're describing performing surgery with a tool designed for moving boulders," Corvin said.

"Yes," she said simply. "That is exactly what I am describing."

Pim, resting on a folded equipment pad with his injured leg extended — the Survey Office physician had confirmed bruised bone, nothing fractured — watched her with steady amber eyes. "You've been preparing for this since the ridge," he said. "I can tell because you have that particular expression."

"What expression?"

"The one that means you've already decided you're going to succeed and you're annoyed it isn't done yet."

Hetha permitted herself a small smile, then straightened and picked up her wand.

She began with the sample in Lys's observation chamber. She pressed the crystal tip of her wand to the glass, closed her eyes, and reached inward to the place where her magic lived — not the broad, sweeping buoyancy force of the Levity Draft or the crushing certainty of the Anchor Press, but something slower and far more precise, a technique she would later record in her compendium simply as the Molecular Flush. She thought about helium at the molecular level: the smallest, lightest noble gas, with no electron bonds to form, perfectly non-reactive, sliding between everything else in the universe with absolute indifference — the same real property that lets welders use helium gas to shield hot metal from the air without the gas ever reacting with the weld itself.

She pushed that concept, that precise energetic intention, through the crystal tip at a scale she had never attempted before.

For a long moment, nothing visible happened.

Then Lys said, in her bell-harmonic voice: "The matrix is responding. I can feel the displacement beginning at the bonding sites. It's — " she paused, her spectral form flickering with concentration — "it's working. Slowly. But it's working."

Corvin's analyzer confirmed it. He watched the readings with an intensity that made him forget to maintain his habitual composure. "Compound concentration in the sample matrix is dropping," he said. "Point by point. Hetha — it's actually dropping."

The effort was immense. Hetha's hand trembled with the sustained precision of it, sweat gathering at her temple. Working at molecular scale with magic designed for the physical world was like trying to thread a needle while wearing heavy gloves — possible, but demanding a quality of focus that left no room for anything else.

After forty minutes, the sample crystal was clean.

Hetha lowered her wand and breathed for a moment, standing still.

"Both nodes," Pim said gently. "You'll need to do both."

"I know," she said. She flexed her hand, rolled her shoulders, and picked up the wand again. "Corvin, connect me to the Caldervane node feed through the pipe network. Tallo, open junction seven and route the atmospheric channel to the primary interface."

Tallo threw the junction lever. Corvin made the connections.

Hetha pressed her wand to the interface console and sent her magic into the mountain.

The Caldervane node cleansing took three hours. The Greyvane node — reached through a secondary interface Corvin jury-rigged from the monitoring cable network — took two more. When it was done, Lys confirmed both with a long, resonant tone of pure relief that made the crystal formations in the lab sing softly in response.

Hetha sat down on the floor with her back against the bench, wand in her lap, compendium beside her, and said nothing for a very long time.

Pim limped over and sat beside her, leaning his warm, rust-orange weight against her arm.

"Well done," he said quietly.

She leaned her head back and looked at the pipe-threaded ceiling. "We still have fifty-four hours to spare," she said.

"I know," he said. "I'm still impressed."

Chapter 10: What Remains After the Storm

Three weeks later, Thornmere was having a festival.

It was not precisely a festival in celebration of the saving of MarBryn's atmospheric node network, because the Survey Office had handled the public communication of those events with the careful understatement of an institution that preferred its crises managed quietly. The official report described "elemental maintenance operations successfully conducted in the northern node corridor," and left it at that. Hetha had not minded. She had not done it for the announcement.

The festival was the annual Alchemist's Lantern Night, a tradition older than the pneumatic rail and newer than the limestone buildings — an evening when the citizens of Thornmere hung new alchemical lanterns outside their homes and businesses, replacing the old ones, and the chemical-gas artisans who made them displayed their newest formulations in the market square. Colors this year ranged from the conventional amber and warm white to an extraordinary new deep violet — achieved by a young gas artisan named Petra who had spent eighteen months on the compound formulation and was receiving, this evening, the recognition she deserved. Petra herself was round-faced and perpetually flour-dusted despite working in chemistry rather than a bakery, with the particular glow of someone whose obsession had finally paid off in front of a crowd, and she kept a candied violet — matched, proudly, to her own lantern color — tucked in her apron pocket for luck.

The market stalls that evening were as much about food as about light. Hetha stood at the edge of the market square with Pim beside her, watching the lanterns go up, a paper cone of roasted chestnuts warming her hands — her one true festival indulgence, since she otherwise ate the way she did everything else: efficiently. Pim had already worked his way through two skewers of honey-glazed meat and was eyeing a third with the patient, unhurried confidence of someone who knew exactly how this evening would end. The square was full of warm light and warm noise — people calling to neighbors across the cobblestones, children chasing each other between the display stalls, the smell of spiced food from the vendors along the eastern row. A steam-calliope at the square's center played something cheerful and slightly out of tune.

Pim's leg had healed cleanly. He stood on all four, his three tails raised in the relaxed, fanned position that indicated genuine contentment.

"Corvin left this morning," Hetha said.

"I know. He said goodbye to me before dawn." Pim's amber eyes were warm. "He has a contract in the southern refinery district. Something about recalibrating a gas chromatography system that's been misreading sulfur concentrations for two years." He paused. "He's a good man. He'll come back through."

"He will," she agreed.

Corvin Strake had spent his last evening in Thornmere at Hetha's laboratory table, drinking tea and reviewing the data from the entire operation with the focused interest of someone who intended to write a very thorough field report. He had shaken hands with Hetha at the door with a firm grip and gray eyes that carried something warmer than professionalism. "Next time you need an engineer in a difficult situation," he had said, "you know where to find me." She had told him she would remember that, and she meant it.

Tallo had departed two days earlier, recalled to the Survey Office in Caldervane for what Director Solley had called "a well-deserved formal commendation and a significant promotion." He had left Thornmere in a state of barely contained excitement, his map case under one arm and his geology hammer cleaned and polished for what was apparently a ceremonial occasion. He had embraced Pim, shaken Hetha's hand, and told Lys that she was the most scientifically remarkable being he had ever encountered, which had made Lys's spectral form flush an extraordinary warm rose color.

Lys was still in Thornmere. That had been the most pleasant surprise of the aftermath.

With both nodes cleansed and stabilized, she had found that her awareness of them — previously requiring her physical presence at the Caldervane site — had expanded. She could sense both nodes from anywhere in the northern range now that their matrices were clean and resonant again. She had taken up residence in a disused crystal deposit in the limestone bluffs above Thornmere, where the mineral environment suited her, and had told Hetha with quiet candor that she had been alone for a very long time and found she preferred company.

She appeared now at Hetha's shoulder, her translucent form taking on the colors of the lantern light around her, which made her look, in this particular setting, like a figure made entirely of celebration. Lys did not eat, not in the ordinary sense — her body sustained itself on ambient light and the resonant hum of nearby crystal, the way a plant might live on sunlight — but that had not stopped Pim from insisting, with great ceremony, that she try a bite of honey cake from the festival stalls. She had accepted out of curiosity more than hunger, and the sweetness had passed through her translucent form in a visible ripple of soft gold light, delighting her so thoroughly that she asked, quite seriously, whether it would be possible to have some sent up to the bluffs on a regular basis.

"The violet lantern compound," Lys said, her bell-harmonic voice carrying a note of admiration, "uses a manganese-based luminescent agent. I can see the molecular structure in its emission spectrum. It's beautiful."

"I'll introduce you to Petra," Hetha said. "She'd appreciate a spectral analysis from a Prism Sprite."

Lys smiled — a rare expression on her fine, light-edged face, and all the more striking for it. "I would like that."

Hetha looked up at the lanterns — the amber ones, the white ones, and the new extraordinary violet — rising on their hooks and brackets above the cobblestone square, their alchemical light steady and clean. She thought about Mordavek in his containment cell in the Hall of Elemental Governance, where the enforcement officers had assured her he would remain for a considerable time, his iron arm dismantled by their technicians, his compound agents catalogued and secured. She thought about the nodes, humming cleanly in their crystal chambers. She thought about the compendium in her satchel, its pages already accumulating new notes from the past three weeks in her cramped handwriting — molecular-scale helium displacement, tri-phase reactive mineral compound classification, Flarkin electrochemical tail discharge observed under field conditions.

She thought about how much there was still to know, and found that thought, as always, deeply pleasing.

"What comes next?" Pim asked, reading her expression with the accuracy of long acquaintance.

Hetha reached into her satchel and withdrew the compendium, opening it to the back pages where she kept her own research notes rather than reference material. She showed him a page she had been working on for the past week — dense with notation, molecular diagrams, and two columns of cross-referenced elemental properties.

"There are six other atmospheric nodes in MarBryn," she said. "Now that I know what Mordavek's compound looks like in a crystal matrix, I want to survey all of them. Make sure none carry residual contamination from earlier operations we didn't know about."

Pim looked at the page, then up at her. "That would take months."

"It would," she agreed. "Interesting months."

He settled his three tails in a slow, deliberate wave of decision. "I find I have no conflicting appointments," he said graciously.

Lys reached out and touched the compendium's open page with one light-refracting finger, reading the notation with those enormous spectral eyes. "The southern nodes pass through crystal formations I have not visited since I was very young," she said. "I would not object to seeing them again."

Hetha closed the compendium, clasped her satchel, and looked out at Thornmere's lantern-bright square — at the steam-calliope playing its cheerful, off-key music, at the violet light of Petra's new compound burning steadily above the market stalls, at the familiar ridgeline of Greyvane above the rooftops, clean and dark against the evening sky.

She felt, as she always did at the beginning of something, that particular mixture of precise readiness and open curiosity that had been her character since childhood — the same quality that had led her to her first compendium at age nine, to her first helium crystal at fourteen, to every laboratory and trail and node chamber since.

She raised her wand briefly, letting the crystal tip catch the lantern light and throw small prismatic sparks across the cobblestones at their feet. Pim watched them scatter with his ears perked and his tails high.

"Then we leave at first light," Hetha said.

And the lanterns of Thornmere burned on, steady and bright, in every color the alchemical arts could make.

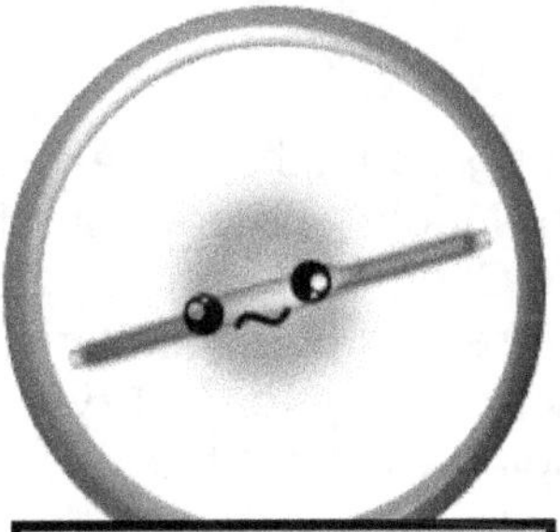

Helium Facts

- Discovered in 1895 in London and Sweden
- Non-reactive to other chemicals
- High thermal conductivity
- Glows pink-orange when electrified
- Noble Gas

Author's Note

Thank you for picking up Hetha the Alchemical Wizard with Helium Magic — the second book in the Magical Elementals Chemistry Fantasy Series. If you met Element 1 before this one, welcome back. If Hetha is your very first stop in MarBryn, welcome in — I think you're going to like it here.

When I sat down to write this book, I knew I wanted an element that felt like a contradiction, because contradictions make the best magic. Helium is the second-lightest element in the universe, and yet real helium is famously "unsociable" — a noble gas that refuses to bond with almost anything else. I loved the idea of building a heroine whose power comes not from force or fire, but from something gentle, buoyant, and quietly unshakeable. Hetha doesn't blast her way through trouble. She lifts, she balances, she presses gently until a problem locks into place instead of exploding. That felt like exactly the kind of magic — and the kind of person — I wanted to spend a book with.

Writing Hetha also meant falling back in love with real chemistry. Every strange word in her compendium — atmospheric nodes, tri-phase reactive compounds, molecular bonding sites, noble gas flushes — is rooted in something true about how elements actually behave. Helium really is chemically inert. Gases really can be used to displace other substances at a molecular level. I wanted young readers (and plenty of grown-up ones, too) to close this book having absorbed a little real science without ever feeling like they sat through a lesson. If Hetha sends anyone down a rabbit hole looking up noble gases or crystal structures after the last page, I will consider this book a wild success.

I also fell for the company Hetha keeps. Pim gave me an excuse to write loyalty with a sense of humor. Corvin let me put engineering and magic side by side as equally valid ways of understanding the world — something I care about deeply. Tallo reminded me that expertise doesn't require magic at all, just curiosity and a good hammer. And Lys — well, Lys surprised me. I didn't know, when I started this book, that I was writing a story about loneliness finding its way toward company. She did that on her own, the way good characters sometimes do.

If this book leaves you looking twice at a birthday balloon, or wondering what other "boring" elements might be hiding something remarkable, then Hetha did exactly what I hoped she would do.

Thank you for reading, for believing a wizard could be a chemist, and for coming along on the Alchemist's Road with Hetha, Pim, and the rest of this strange and wonderful company.

With gratitude and a little bit of helium,

Sybrina Durant

PS: If you're interested in reading the resources the scientific data in this book was derived from, check out this webpage:
https://magicalptelements.com/mw1pt/

About the Author

Sybrina Durant is the author of

"The Magical Elementals Chemistry Fantasy Series"

"The Magical Elements of the Periodic Table Series"

"The No Metal No Magic Series"

"The Blue Unicorn – Journey To Osm" books

and she is the creator of the "Elemental Fantasy Periodic Table" which is available as a brilliantly colored poster.

There are also book activities, songs, trading cards and t-shirts for all 118 elements and elementals, too.

Visit MagicalPTElements.com to see everything that is currently available.

If you love to sing and learn, check out the "Chemistry Sing Alongs" playlist on the Sybrinad channel on Youtube.

Glossary of Scientific Terms in Hetha the Alchemical Wizard with Helium Magic

Helium (He)

Real Science: Helium is the second-lightest element in the universe and a noble gas, meaning it almost never forms chemical bonds with other elements under normal conditions. It is about seven times less dense than air, which is why helium-filled balloons rise, and it is used in the real world for everything from party balloons to cooling powerful magnets in MRI machines and particle accelerators, precisely because it stays chemically inert even under extreme conditions.

Story Magic: In Hetha's hands, a crystal of solidified helium sealed into her wand becomes a direct channel for her will. She can push the helium's real lightness outward as buoyant lift (the Levity Draft), or reverse the flow to add crushing density (the Anchor Press) — and, at her most advanced, project helium's non-reactivity down to the molecular level to physically clear away a dangerous compound (the Molecular Flush).

Noble Gas

Real Science: Noble gases — helium, neon, argon, krypton, xenon, and radon — sit in the rightmost column of the periodic table because their outer electron shells are already full, which leaves them with almost no motivation to bond with other atoms. This is why they are called "inert" or "noble": they mostly keep to themselves.

Story Magic: Hetha's entire magical discipline is built on this one chemical fact. Because helium refuses to react with what it touches, her spells can move through solid stone, living creatures, and reactive chemical compounds alike without ever triggering an unwanted reaction of their own — which is exactly what lets her later use a "noble gas flush" to cure the nodes without setting off a second explosion.

Buoyancy

Real Science: Buoyancy is the upward force a fluid (including a gas like air) exerts on anything less dense than itself. A cork floats on water and a helium balloon floats in air for the same underlying reason: each is lighter than the fluid it displaces.

Story Magic: The Levity Draft is Hetha's name for buoyancy turned into a directed spell. Instead of filling a balloon, she convinces the air around a person or object to behave, briefly, as though it were far lighter than usual, letting real buoyant force do the actual lifting — which is why the spell feels, to Pim, like "a breath drawn in" rather than an explosion of power.

Density

Real Science: Density describes how much mass is packed into a given volume. Compressing a gas into a smaller space raises both its density and its pressure — the same reason a scuba tank feels far heavier than the air inside it would suggest, and the same reason deep water presses harder on a diver than shallow water does.

Story Magic: The Anchor Press is Hetha's reverse spell: instead of lightening something, she gathers density into a tight, heavy column around it. It is this technique that lets her lock Mordavek's trigger devices — pressing them down past their activation threshold instead of letting them fire upward — and later lift and pin Mordavek himself with the Column Lift.

Molecular Bonding Sites

Real Science: At the microscopic level, atoms and molecules attach to one another at specific "bonding sites" within a material's structure — the exact points where chemical forces hold a substance together. Removing a contaminant from a material often means physically or chemically clearing it away from these sites.

Story Magic: When Hetha performs the Molecular Flush in the Academy's node interface laboratory, she is pushing her usual broad, physical magic down to this microscopic scale for the first time — using helium's harmless inertness to nudge Mordavek's reactive compound molecules off the crystal's bonding sites without triggering the very reaction she's trying to stop.

Gas Purging / Inert Gas Flush

Real Science: In real engineering, inert gases like helium or nitrogen are used to "purge" or "flush" a system — pipelines, welding sites, sealed containers — displacing oxygen or other reactive substances without themselves reacting with anything in the process. Welders, for instance, often shield hot metal with a stream of helium or argon so the weld doesn't react with the surrounding air.

Story Magic: This is the real-world technique the compendium's cure for Mordavek's compound is based on: a "noble gas flush," delivered at enough pressure to physically displace the reactive tri-phase compound from the crystal matrices at Greyvane and Caldervane, is what makes the Molecular Flush possible.

Atmospheric Nodes

Real Science: There is no real-world equivalent of a magical atmospheric node, but the idea borrows from how real pressure-regulating valves work in gas pipelines — controlling flow, pressure, and distribution across a connected network so that no single point becomes dangerously overloaded.

Story Magic: In MarBryn, atmospheric nodes are natural crystal formations that perform this same regulating function for elemental energy across the land. Mordavek's plan to corrupt both the Greyvane and Caldervane nodes simultaneously is built on treating them like two valves in the same pipeline — destabilize both, and the pressure differential between them becomes a weapon.

Pneumatic Systems

Real Science: "Pneumatic" describes any system or device powered by compressed or pressurized gas, usually air. Real pneumatic systems include air brakes, pipeline pigs, jackhammers, and industrial lift systems — anywhere stored gas pressure is released to do mechanical work.

Story Magic: Thornmere and Caldervane's "new-world" technology — forges, gates, Corvin's collapsible scope, the abandoned summit lift track — all runs on this same real principle. It gives the world a grounded, technological cousin to Hetha's magic: both are, at their core, about directing pressurized gas to do useful work.

Catalytic Agent / Trigger

Real Science: A catalyst is a substance that speeds up or initiates a chemical reaction without being consumed by it. A trigger device, in engineering terms, is simply a mechanism designed to release stored energy once a specific condition is met — such as a pressure threshold.

Story Magic: Mordavek's trigger devices at Greyvane and Caldervane combine both ideas: a catalytic compound paired with a pressure-sensitive mechanical trigger. Hetha's Anchor Press defeats them by exploiting the same pressure-sensitivity — driving density downward instead of allowing the upward pressure the mechanism needs to fire.

Volatile Compound

Real Science: A volatile compound is one that is chemically unstable and prone to sudden, often dangerous changes in energy or pressure — real-world examples include many industrial solvents and unstable chemical mixtures that must be stored and handled with great care.

Story Magic: The wrong-colored green light that opens the story is the visible signature of volatile compound magic, and it stands as the clearest opposite to Hetha's own craft: where Mordavek's chemistry is deliberately reactive and destabilizing, Hetha's is built on an element that refuses to react at all.

Tri-Phase Reactive Mineral Compound

Real Science: While "tri-phase reactive mineral compound" is an invented name, it borrows the real concept of a self-propagating reaction — a chemical process that, once it reaches a certain concentration (a "saturation threshold"), continues on its own without needing further outside energy, the way a fire can become self-sustaining once enough fuel is burning.

Story Magic: This is the compound Mordavek seeds into both node matrices. Once it reaches saturation, it will trigger passively, with no mechanical device required — which is why locking the trigger devices only buys Hetha and her companions time, not a solution, and why the Molecular Flush becomes necessary.

Spectral Analysis

Real Science: Spectral analysis is a real scientific method of reading the specific wavelengths of light absorbed or emitted by a substance to identify its chemical composition — the same principle astronomers use to determine what distant stars are made of, and chemists use to verify unknown samples.

Story Magic: Caldervane's customs officers use spectral analysis on Corvin's gas unit to confirm it is safe, and Lys — a being literally made of living light — performs a version of the same trick instinctively, reading a crystal's or a lantern's molecular structure directly from the light it gives off.

Static Discharge (Triboelectric Effect)

Real Science: Rubbing two materials together can transfer electrons from one surface to the other, building up a static electric charge that discharges suddenly when it finds a path — the same everyday effect behind a shock from a doorknob or a sock pulled fresh from the dryer.

Story Magic: A Flarkin's three tails build up exactly this kind of charge through the natural friction of their thick fur. Pim's Static Snap is that stored charge released in one controlled burst — a small, sharp jolt that can knock a weapon loose or scramble a mechanical arm's calibration.

Refraction and Optical Dazzle

Real Science: Refraction is the bending of light as it passes through a material like glass, water, or crystal, splitting white light into its full spectrum of colors — the same principle that makes a prism throw a rainbow, or a diamond sparkle from every angle.

Story Magic: Lys's Refractive Bloom takes this real property and weaponizes it, splitting and scattering light violently enough, and fast enough, to overwhelm an opponent's eyes with more color and motion than they can process in an instant.

Acoustic Reflection (Sound Redirection)

Real Science: Sound waves, like light waves, can be reflected and redirected by hard, resonant surfaces — the same principle behind a whispering gallery, where a sound spoken quietly at one point in a curved room can be heard clearly at another point far away.

Story Magic: Lys's Prism Throw bends sound the way a curved mirror bends light, bouncing a resonant thrum off the node's crystal formations so it seems to come from a direction she isn't actually standing in — the trick that buys Hetha the three seconds she needs at the Greyvane arch.

Corrosive Agent / Acid Erosion

Real Science: Acids and other corrosive chemicals can break down the molecular bonds holding rock, metal, or other materials together over time — real-world acid rain, for instance, slowly erodes limestone and marble structures through exactly this kind of chemical reaction.

Story Magic: Mordavek treats the summit ledge with a corrosive agent timed to degrade under foot pressure, and later fires similar compounds from his arm-mounted casing. It is the clearest chemical opposite of Hetha's inert helium magic in the entire book — reactive destruction versus stable, noble restraint.

Calcium Carbonate / Limestone and Metamorphic Rock

Real Science: Calcium carbonate is the mineral compound that makes up limestone, formed over millions of years from compressed marine sediment. Under enough heat and pressure, limestone can transform into a metamorphic rock like marble — a real geological process Tallo narrates almost word for word as they climb.

Story Magic: These are the literal mineral strata the atmospheric nodes are embedded in. Understanding the rock, as Tallo does, is part of understanding the physical "container" that Hetha's magic has to work inside during the Molecular Flush.

Gas Chromatography System

Real Science: Gas chromatography is a real analytical technique used to separate and measure the exact concentration of different gases or vaporized compounds within a mixture, widely used in industrial safety monitoring, environmental testing, and chemical research.

Story Magic: Corvin's post-story contract — recalibrating a gas chromatography system in the southern refinery district — is presented as the "new-world," instrument-based cousin to Hetha's "old-world" magic: both, in the end, are in the business of understanding what gases are present and what to do about them.

Luminescent Agent (Manganese-Based)

Real Science: Certain metal compounds, including manganese-based ones, are used in real-world luminescent and phosphorescent materials to produce specific colors of glowing light, from fireworks to glow-in-the-dark paints.

Story Magic: Petra's award-winning violet alchemical lantern, unveiled at the Lantern Night festival, uses exactly this kind of compound chemistry — a small, celebratory echo of the same faint luminescence given off by Hetha's own helium crystal.

www.ingramcontent.com/pod-product-compliance
Lightning Source LLC
La Vergne TN
LVHW050747061126
851857LV00013B/1036
* 9 7 8 1 9 4 2 7 4 0 7 1 1 *